Collard, Sneed B.

Our natural homes.

DATE			

2/97

Our Natural Homes

Exploring Terrestrial Biomes of North and South America

Sneed B. Collard III
Illustrated by James M. Needham

Charlesbridge

To Maryam, for helping me plant trees
—S. B. C. III

For my wife and best friend, Charlene, and for my large and loving family, who keep my ship in safe harbor
—J. M. N.

Published by Charlesbridge Publishing
85 Main Street, Watertown, MA 02172-4411
(617) 926-0329

Library of Congress Cataloging-in-Publication Data
Collard, Sneed B.
Our natural homes: exploring terrestrial biomes of North and South America / by Sneed B. Collard III; illustrated by James M. Needham.
p. cm.—(Our perfect planet)
Summary: Defines the earth's land ecosystems through the characteristic plants and animals found in each.
ISBN 0-88106-929-9 (trade hardcover)
ISBN 0-88106-930-2 (library reinforced)
ISBN 0-88106-928-0 (softcover)
1. Habitat (Ecology)—Juvenile literature. [1. Habitat (Ecology).]
I. Needham, James, ill. II. Title. III. Series.
QH541.14.C65 1996
574.5'26—dc20 95-23978

Printed in China
(hc) 10 9 8 7 6 5 4 3 2 1
(sc) 10 9 8 7 6 5 4 3 2 1

The paintings in this book are done in gouache on Crescent illustration board.
The display type and text type were set in Usherwood.
Color separations by Pure Imaging, Watertown, Massachusetts
Printed and bound by Palace Press International
Production supervision by Brian G. Walker
Designed by Diane M. Earley

EARTH'S BIOMES

Where do we live? It sounds like a simple question. We live in houses and apartments. We live on streets, and in states and countries, too. But each of us also lives in a natural home—a home that was here long before people were. Scientists have a name for our natural homes. They are called *biomes*.

Biomes can be on land or in water. This book takes you through the land, or *terrestrial,* biomes of North and South America. Figuring out what biome you are in isn't always easy. People have changed most biomes by building cities, planting crops, and cutting down trees.

Look for the right trees, animals, weather, and other clues, however, and you can still discover the biomes of our planet. To start collecting clues about earth's different natural homes, put on your hiking shoes and grab your backpack. Where do we live? Let's take a trip to find out.

TUNDRA

If you like cold—and I mean *cold*—weather, you'll love this biome. During the long, dark winter, temperatures plunge to seventy degrees below zero. Legions of lemmings survive by burrowing under the snow and gobbling up buried lichens, mosses, and shrubs. That shaggy white rock is a polar bear with special fur and a thick layer of fat to keep her snug.

When summer finally comes, the sun shines twenty-four hours a day, and the snow melts into shimmering ponds and streams. Geese and sandpipers migrate here to feast on swarms of mosquitoes and crane flies. Caribou come to graze and give birth to their young. Most visitors don't stay long. In a few short weeks, winter will again throw its cloak over the tundra, driving them away for another long, cold season.

BOREAL FOREST or TAIGA

Just south of the tundra, a four-thousand-mile-wide boreal forest, or *taiga,* stretches from Newfoundland all the way to Alaska. Spruce are the most common trees here. Often, they are packed so close together they block out light for other plants. Most spruce don't grow very tall because taiga soils are poor, and long, cold winters keep the growing season short.

Pictures of the taiga often show moose. Moose live in many places, but this is their favorite home. Moose weigh up to one thousand pounds. In winter, their four stomachs help them digest twigs and bark. In summer, moose eat fresh leaves and dunk their heads underwater to reach tasty water weeds. As in the tundra, millions of birds migrate to the taiga each summer to raise new families. Clear the runway! There's a rare whooping crane coming in for a landing.

MOUNTAINS

The beauty of mountains can take your breath away. So can the cool air. This biome is way up high, where air is thin and it's hard to breathe. The sun doesn't feel hot, but it will burn bare skin. Only the hardiest plants survive up here. Pine and spruce trees cling to rocks and thin layers of soil, but they are stunted by the cold and the wind. Very high up, you're lucky to find even a patch of moss or lichens.

During winter, snow falls up to thirty feet deep here. Pikas gather hay-piles of plants to feed them through the winter. Most other animals hibernate or spend the coldest months down in the warmer valleys. In spring, snow melts, and columbine and goldenrod flowers turn mountain slopes into dazzling meadows. That grizzly bear has just left its winter den. Right now it's slurping up juicy earthworms, but keep your distance — it can chase down a speeding mountain goat when it's in the mood.

TEMPERATE DECIDUOUS FOREST

Each fall, this biome blazes with color as maples, oaks, and birches drop over ten million leaves on every acre of rich forest soil. These trees are called *deciduous* trees because they lose their leaves each year. Poke under the crunchy leaf layer and you'll find more salamanders than anywhere else in the world. Their colors may make them look like toys, but don't touch. Their bright skins warn: "I have poisonous skin. Play with someone else!"

By dropping leaves, deciduous trees protect their tender tissues from the oncoming winter. Still, this biome is a *temperate* place— it rarely gets too hot or too cold. Winter snows soon turn to spring rains as symphonies of songbirds migrate here from the tropics. Through the summer, the birds eat insects and raise new babies. Then they return south as the lovely rain of fall leaves begins once more.

TEMPERATE EVERGREEN RAIN FOREST

If you want to hug a giant, don't miss this biome. Everywhere you look, three-hundred-foot redwood and fir trees tower over you. They sprout from rich soil and are watered by up to fifteen *feet* of rainfall each year! Like the spruce and pine trees you've already seen, most trees here are *evergreen.* They hang on to their leaves or needles all year long and keep the forest smelling spicy and fresh.

Standing in this forest is like being in a living library. You don't hear much, but a collection of life surrounds you. Elk keep the forest floor mowed by grazing on ferns, skunk cabbage, and liverworts. Berries abound—if you pick them, make sure a black bear doesn't join you! Rotting logs provide food and shelter for almost two hundred other kinds of birds, mammals, and amphibians—not to mention that banana slug crawling up the tree.

TEMPERATE GRASSLANDS or PRAIRIE

Temperate grasslands are "where the deer and the antelope play." In every direction, blue grama and buffalo grass stretch beyond the horizon. These plants are great grub for buffalo and other big grazing mammals. Look closely—the grasses are also home to restless armies of crickets, butterflies, and spiders.

Only about ten inches of rain falls on these grasslands each year. Drying winds never cease. Prairie dogs, jumping mice, and other rodents dig tunnels into the earth. Rattlesnakes and black-footed ferrets also crawl into these cool, dark passages to hunt. In tunnels, animals need less water than they do above ground. A tunnel is the best place to be when temperatures soar and summer lightning-fires sweep across the prairie.

DESERT

Deserts are even drier than grasslands. Here, the sun scorches the earth and rain rarely falls. Cactuses and many other plants have thick leaves that store water and waxy skins that keep them from drying out. Handle with care. Cactuses are also covered with sharp spines. These spines help block the sun's rays and keep the cactuses from "cooking."

Billions of beetles and other insects thrive in the desert. Their hard shells, or *exoskeletons,* help their bodies hold on to water. Someone might tell you that the venomous Gila monster can kill you by breathing on you. Don't believe it! These lazy lizards must bite you to inject their poison — and they will do that only to defend themselves. Like most animals here, Gila monsters stay underground or in the shade during the hottest days. Under the cool moon, they venture out to hunt, mate, and explore.

CHAPARRAL

Chaparral plants grow mostly in winter—not in summer. That is because the sun shines all year, but only in winter do storms bring rain so plants can grow. Oak trees and sweet-smelling shrubs hold the soil in place on this biome's rocky slopes. You may have seen these plants in Hollywood movies. However, the bigger "stars" here are the endangered California condors—soaring overhead on their nine-foot wings.

Like many desert plants, chaparral plants have thick, fleshy leaves that help them survive hot, dry weather. In summer and fall, blistering Santa Ana winds whip roaring fires over the land. The blazes often terrify people, but without the fires, many plants would disappear. A lot of chaparral seeds need fire to sprout. Fires also burn away old, dead brush and let in light for new plants to grow.

WARM TEMPERATE FOREST

In this biome, year-round sunshine keeps bodies warm, and frequent rains water a cornucopia of pleasing plants. On drier high ground, long-leaf pines reach toward the sky. Gopher tortoises graze on grass and dig thirty-five-foot-long burrows into the sandy soil. An assortment of creatures, from indigo snakes to Florida mice, share these natural community centers.

As you walk down into the swampier areas — or *bottomlands* — of this biome, you'll find that pines are replaced by tupelo, magnolia, and cypress trees. These plants put up with months of flooding as water from winter and spring storms covers wide, shallow valleys. Over nine-tenths of the birds in eastern North America spend at least part of the year in these forests. That night heron better watch out for the big, bumpy log floating nearby!

SAVANNA or TROPICAL GRASSLANDS

Temperate biomes have four seasons. Here in the savanna, there are only two — a *wet season* and a *dry season*. During the wet season, frequent rains pound the soil. The rest of the year — the dry season — the earth is baked hard by the hot tropical sun. Savanna soils are poor in nutrients, but hundreds of species of bunch grasses, woody shrubs, and other flowering plants grow here. In places, scattered trees also cast precious shade.

Near the end of the dry season, fires sweep across these grasslands. The fires burn away dry stems and release nutrients for new plant growth. Grasses and herbs quickly sprout after a fire. Twenty thousand years ago you might have seen mammoths, giant ground sloths, and pony-sized horses eating the plants. Now, deer and the world's largest rodents — capybara — grow fat on these natural pastures.

TROPICAL DRY FOREST

Like the savanna, the tropical dry forest also has only two seasons. In the wet season, swarms of insects make juicy snacks for ctenosaur lizards and trogons raising their young. In a tree full of figs, you're likely to see a family of white-faced monkeys. The monkeys help spread the forest seeds around, but watch your head—they also have a good aim with ripe figs!

When the dry season begins, a host of deciduous trees drop their leaves. This helps them use less water. Many plants also flower now. This is good news for hummingbirds and bees, but for most animals the dry season is a tough time. Monkeys, deer, and tapirs gather near streams, where there's more food and water. Butterflies migrate to wetter forests or become inactive until the rains and food return.

TROPICAL RAIN FOREST

To finish your journey, you'd better pull on some rain boots. In this biome, there's only one season — wet! In the tropical rain forest, rain falls in buckets every month of the year. The storms water so many plants that it's often hard to see the sky. On just a couple acres of forest, you can find more kinds of trees than grow in all of the United States and Canada.

More species of animals live here than anywhere else on earth. Biologists believe that thirty to fifty *million* kinds of insects fly, crawl, and burrow through this forest. They share this biome with thousands of species of reptiles, amphibians, and birds. Do you see that jaguar hiding behind the tree? Jaguars are the largest wildcats in North and South America and the top hunters here. These big cats are shy and stay well away from people.

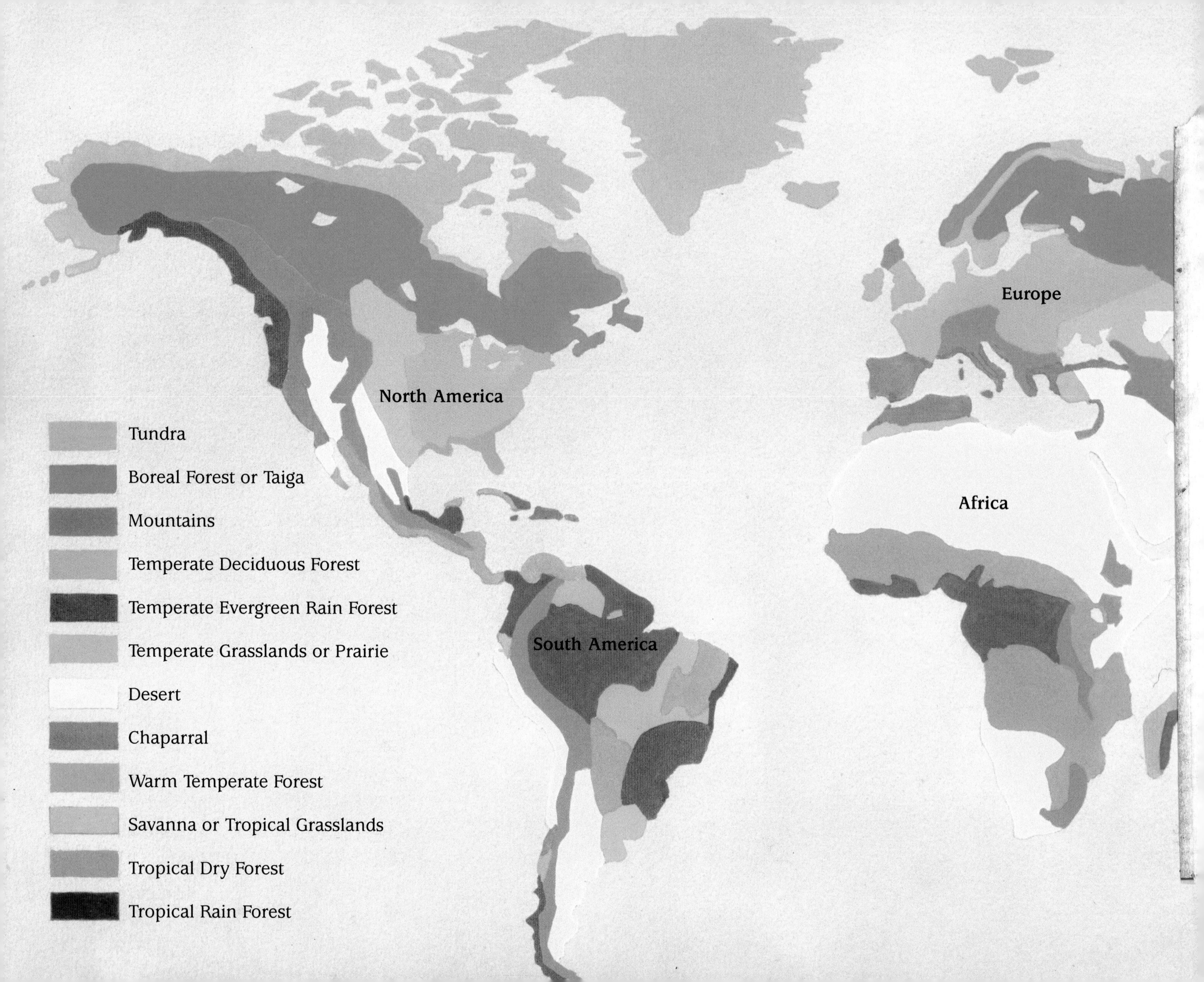

North America
Europe
Africa
South America
Tundra
Boreal Forest or Taiga
Mountains
Temperate Deciduous Forest
Temperate Evergreen Rain Forest
Temperate Grasslands or Prairie
Desert
Chaparral
Warm Temperate Forest
Savanna or Tropical Grasslands
Tropical Dry Forest
Tropical Rain Forest

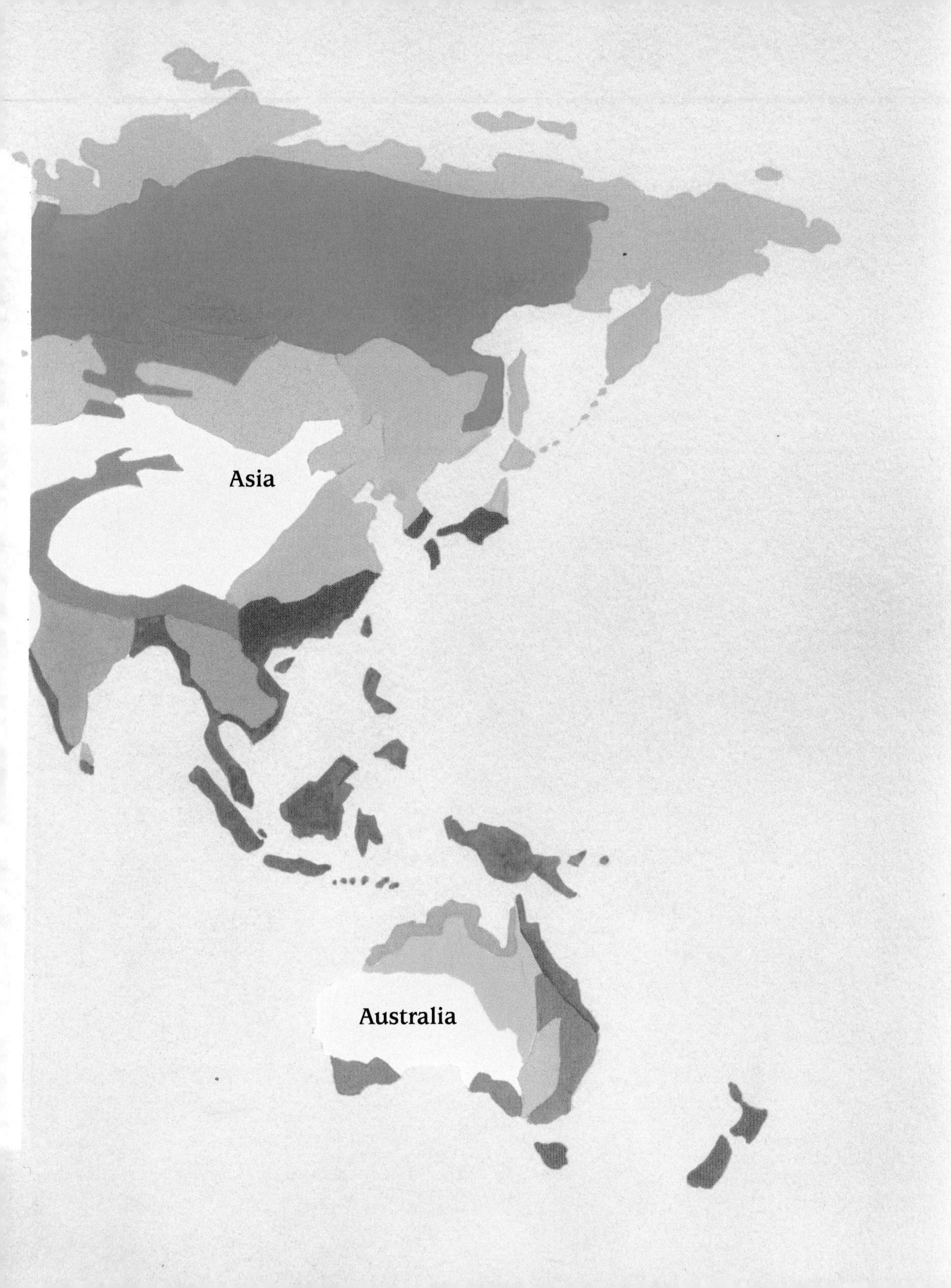

MANY BIOMES— ONE PLANET

Congratulations! You've just traveled through every major biome in North and South America. As you can see from the map, many of the biomes you've visited are also found in other parts of the world. The plants and animals may look different in different places. However, the basic weather, soils, and seasons of our natural homes are the same in America as in Asia, Europe, Africa, or Australia.

It's important to learn about all of earth's natural homes, because all biomes work together to support life on earth. Forests make oxygen that all of us use. Mountains collect snowfall that melts to water places thousands of miles away. Visiting different biomes is fun, but it also helps us make smart choices—choices about taking care of the wonderful planet that we all depend on. After all, there's really only one answer to the question, "Where do we live?" We *all* live on . . . **planet earth!**

GLOSSARY

amphibians: the group of cold-blooded, vertebrate animals that includes frogs, toads, and salamanders. Most amphibians spend at least part of their lives in water. Young amphibians usually breathe through gills underwater and develop air-breathing lungs as they become adults.

biome: a large natural community or "neighborhood" of plants, animals, and other life-forms. This term also refers to the weather and physical landscape where a community is found.

bottomlands: parts of the warm temperate forest that lie along rivers and valleys. Bottomlands are flooded during much of the year and are often called swamps.

deciduous: trees and other plants that drop their leaves each year.

dry season: the part of the year when rain is rare. Depending on where you are, the dry season can be as short as a month or last most of the year.

evergreen: trees and other plants that keep their leaves all year long. Most pine trees are evergreen.

exoskeleton: a skeleton that is on the outside of an animal instead of on the inside. The hard shells of insects and crabs are exoskeletons.

hibernation: a deep resting, or dormant, state that allows animals to use very little energy. Animals usually hibernate in extremely cold or hot weather or when there are shortages of food or water.

lichens: small, plantlike structures that often grow in places that are very cold and dry. A lichen is made up of two living things growing together, a fungus and an alga.

liverworts: small green plants that are related to mosses. They look like little shingles and usually grow in wet forests.

mammals: the group of warm-blooded animals that includes humans, cats, bears, lions, and rodents. Female mammals breast-feed their young, and most mammals have hair or fur.

migrate: to travel long distances regularly. Each summer, caribou migrate north to the tundra to find food.

nutrients: chemicals such as nitrogen and phosphorus that plants and other living things need to survive and grow.

pikas: small, rabbitlike mammals that usually live in mountain regions. Unlike many mountain animals, pikas store food and stay active even during the coldest winter weather.

Santa Ana winds: hot, dry winds that blow from the deserts to the coast in Southern California and other places. They greatly increase the danger of wildfires.

temperate zone: a place where temperatures rarely get extremely hot or extremely cold. Most of the United States lies within the temperate zone.

terrestrial: something that lives or exists on land. Wolves are terrestrial animals; whales are not.

tropical: parts of the earth that lie close to the equator. Tropical regions receive more sunlight and are usually warmer than temperate places.

wet season: the part of the year when most of the rain falls. Like the dry season, the wet season can be very long or very short.